HANDWRITING: NUMBER TRACING

PRINTING WORKBOOK, KIDS AGES 3-5

Book 2

Brighter Hand

https://tinyurl.com/tracingletterforkidsages3-5

COPYRIGHT NOTICE

Copyright © 2017 by **HANDWRITING: NUMBER TRACING PRINTING WORKBOOK, KIDS AGES 3-5**.

All rights reserved. This book or any portion thereof may not be reproduced or used in any manner whatsoever without the express written permission of the publisher except for the use of brief quotations in a book review.

CONTENTS

INTRODUCTION 1
HOW TO USE THIS BOOK 2
NUMBERS 1-50 EXERCISES 3 TO 92
CHECK OUT MORE BOOKS BELOW **95**

INTRODUCTION

A child who learns to trace letters, numbers at home, at the early age with their loving parent or caregiver, grows in self-confidence and independence. This promotes greater maturity, increases discipline and lays the basis for moral literacy. A child who begins with early learning books has a distinct advantage over his or her peers. One of the big advantages being there is no psychological pressure. The reason why parent understand why early learning is important..

HOW TO USE THIS BOOK

Practice, Practice, Practice makes life easier and worthwhile so train your child analytical mind by tracing letters and numbers in the alphabet the conventional way through handwritings as the saying said: "**Young children need writing to help them learn about reading, they need reading to help them learn about writing; and they need oral language to help them learn about both.**"

Make your tracing at your own style and convenience all individuals has its own uniqueness. *

NUMBERS 1-50 EXERCISE 1

42   Forty two
45   Forty five
20   Twenty
13   Thirteen
12   Twelve
28   Twenty eight
46   Forty six
17   Seventeen
15   Fifteen
38   Thirty eight
2   Two
29   Twenty nine

NUMBERS 1-50 EXERCISE 2

25	Twenty five
46	Forty six
4	Four
48	Forty eight
5	Five
28	Twenty eight
1	One
39	Thirty nine
50	Fifty
7	Seven
36	Thirty six
31	Thirty one

NUMBERS 1-50 EXERCISE 3

28	Twenty eight
15	Fifteen
23	Twenty three
34	Thirty four
49	Forty nine
26	Twenty six
6	Six
7	Seven
2	Two
9	Nine

NUMBERS 1-50 EXERCISE 4

37	Thirty seven
5	Five
48	Forty eight
49	Forty nine
40	Forty
41	Forty one
19	Nineteen
50	Fifty
16	Sixteen
11	Eleven
35	Thirty five
34	Thirty four

NUMBERS 1-50 EXERCISE 5

39	Thirty nine
6	Six
15	Fifteen
17	Seventeen
50	Fifty
26	Twenty six
40	Forty
31	Thirty one
7	Seven
35	Thirty five
42	Forty two
29	Twenty nine

NUMBERS 1-50 EXERCISE 6

47	Forty seven
3	Three
2	Two
27	Twenty seven
16	Sixteen
24	Twenty four
9	Nine
36	Thirty six
8	Eight
32	Thirty two
6	Six
45	Forty five

NUMBERS 1-50 EXERCISE 7

47 Forty seven
26 Twenty six
5   Five
27 Twenty seven
6   Six
12 Twelve
13 Thirteen
24 Twenty four
44 Forty four
21 Twenty one
43 Forty three
20 Twenty

NUMBERS 1-50 EXERCISE 8

3	Three
45	Forty five
29	Twenty nine
19	Nineteen
7	Seven
15	Fifteen
40	Forty
35	Thirty five
33	Thirty three
9	Nine
4	Four
2	Two

NUMBERS 1-50 EXERCISE 9

34	Thirty four
48	Forty eight
19	Nineteen
12	Twelve
20	Twenty
10	Ten
33	Thirty three
30	Thirty
49	Forty nine
39	Thirty nine
9	Nine
22	Twenty two

NUMBERS 1-50 EXERCISE 10

29	Twenty nine
48	Forty eight
23	Twenty three
47	Forty seven
32	Thirty two
33	Thirty three
3	Three
17	Seventeen
10	Ten
1	One
46	Forty six
4	Four

NUMBERS 1-50 EXERCISE 11

17   Seventeen

40   Forty

9   Nine

47   Forty seven

22   Twenty two

19   Nineteen

12   Twelve

35   Thirty five

26   Twenty six

10   Ten

36   Thirty six

32   Thirty two

NUMBERS 1-50 EXERCISE 12

46	Forty six
27	Twenty seven
47	Forty seven
35	Thirty five
24	Twenty four
7	Seven
33	Thirty three
14	Fourteen
39	Thirty nine
18	Eighteen
49	Forty nine
21	Twenty one

NUMBERS 1-50 EXERCISE 13

39	Thirty nine
44	Forty four
13	Thirteen
8	Eight
14	Fourteen
27	Twenty seven
31	Thirty one
41	Forty one
50	Fifty
28	Twenty eight
12	Twelve
4	Four

NUMBERS 1-50 EXERCISE 14

7	Seven
14	Fourteen
32	Thirty two
22	Twenty two
41	Forty one
2	Two
12	Twelve
42	Forty two
46	Forty six
1	One
33	Thirty three
5	Five

NUMBERS 1-50 EXERCISE 15

44	Forty four
17	Seventeen
40	Forty
22	Twenty two
7	Seven
6	Six
30	Thirty
13	Thirteen
4	Four
25	Twenty five
5	Five
18	Eighteen

NUMBERS 1-50 EXERCISE 16

13	Thirteen
11	Eleven
15	Fifteen
31	Thirty one
34	Thirty four
21	Twenty one
35	Thirty five
47	Forty seven
19	Nineteen
1	One

NUMBERS 1-50 EXERCISE 17

31	Thirty one
25	Twenty five
15	Fifteen
29	Twenty nine
46	Forty six
40	Forty
5	Five
17	Seventeen
13	Thirteen
3	Three

NUMBERS 1-50 EXERCISE 18

23	Twenty three
34	Thirty four
4	Four
15	Fifteen
13	Thirteen
38	Thirty eight
27	Twenty seven
2	Two
36	Thirty six
49	Forty nine

NUMBERS 1-50 EXERCISE 19

45 Forty five

13 Thirteen

41 Forty one

34 Thirty four

9 Nine

17 Seventeen

50 Fifty

7 Seven

29 Twenty nine

49 Forty nine

NUMBERS 1-50 EXERCISE 20

1	One
35	Thirty five
37	Thirty seven
28	Twenty eight
31	Thirty one
27	Twenty seven
25	Twenty five
34	Thirty four
16	Sixteen
30	Thirty

NUMBERS 1-50 EXERCISE 21

3   Three

44   Forty four

16   Sixteen

26   Twenty six

35   Thirty five

25   Twenty five

50   Fifty

33   Thirty three

12   Twelve

13   Thirteen

NUMBERS 1-50 EXERCISE 22

23	Twenty three
28	Twenty eight
22	Twenty two
14	Fourteen
25	Twenty five
2	Two
46	Forty six
33	Thirty three
35	Thirty five
50	Fifty

NUMBERS 1-50 EXERCISE 23

16 Sixteen
13 Thirteen
18 Eighteen
35 Thirty five
19 Nineteen
12 Twelve
7 Seven
26 Twenty six
43 Forty three
33 Thirty three

NUMBERS 1-50 EXERCISE 24

46 Forty six
14 Fourteen
28 Twenty eight
8 Eight
35 Thirty five
50 Fifty
47 Forty seven
15 Fifteen
30 Thirty
43 Forty three

NUMBERS 1-50 EXERCISE 25

8   Eight
19   Nineteen
30   Thirty
27   Twenty seven
20   Twenty
25   Twenty five
1   One
5   Five
36   Thirty six
9   Nine
23   Twenty three
3   Three

NUMBERS 1-50 EXERCISE 26

4	Four
50	Fifty
35	Thirty five
10	Ten
8	Eight
36	Thirty six
44	Forty four
16	Sixteen
3	Three
31	Thirty one

NUMBERS 1-50 EXERCISE 27

49	Forty nine
43	Forty three
16	Sixteen
27	Twenty seven
10	Ten
35	Thirty five
20	Twenty
32	Thirty two
9	Nine
40	Forty

NUMBERS 1-50 EXERCISE 28

46 Forty six
48 Forty eight
43 Forty three
10 Ten
21 Twenty one
39 Thirty nine
32 Thirty two
23 Twenty three
11 Eleven
12 Twelve

NUMBERS 1-50 EXERCISE 29

23	Twenty three
36	Thirty six
5	Five
39	Thirty nine
38	Thirty eight
15	Fifteen
3	Three
18	Eighteen
19	Nineteen
28	Twenty eight

NUMBERS 1-50 EXERCISE 30

40 Forty
28 Twenty eight
19 Nineteen
18 Eighteen
5 Five
50 Fifty
16 Sixteen
44 Forty four
36 Thirty six
10 Ten

NUMBERS 1-50 EXERCISE 31

8   Eight

5   Five

13   Thirteen

11   Eleven

3   Three

4   Four

17   Seventeen

39   Thirty nine

10   Ten

23   Twenty three

NUMBERS 1-50 EXERCISE 32

12	Twelve
25	Twenty five
32	Thirty two
39	Thirty nine
33	Thirty three
7	Seven
27	Twenty seven
14	Fourteen
42	Forty two
46	Forty six

NUMBERS 1-50 EXERCISE 33

24 Twenty four
50 Fifty
17 Seventeen
26 Twenty six
28 Twenty eight
3 Three
23 Twenty three
30 Thirty
16 Sixteen
10 Ten

NUMBERS 1-50 EXERCISE 34

19	Nineteen
50	Fifty
47	Forty seven
4	Four
25	Twenty five
31	Thirty one
16	Sixteen
30	Thirty
15	Fifteen
20	Twenty

NUMBERS 1-50 EXERCISE 35

21	Twenty one
13	Thirteen
9	Nine
40	Forty
36	Thirty six
43	Forty three
6	Six
49	Forty nine
23	Twenty three
48	Forty eight

NUMBERS 1-50 EXERCISE 36

4	Four
49	Forty nine
5	Five
44	Forty four
6	Six
46	Forty six
16	Sixteen
25	Twenty five
42	Forty two
13	Thirteen
2	Two
19	Nineteen

NUMBERS 1-50 EXERCISE 37

50	Fifty
8	Eight
24	Twenty four
40	Forty
47	Forty seven
26	Twenty six
9	Nine
30	Thirty
4	Four
32	Thirty two

NUMBERS 1-50 EXERCISE 38

24 Twenty four
33 Thirty three
4 Four
32 Thirty two
22 Twenty two
39 Thirty nine
44 Forty four
29 Twenty nine
15 Fifteen
37 Thirty seven

NUMBERS 1-50 EXERCISE 39

38	Thirty eight
49	Forty nine
15	Fifteen
45	Forty five
9	Nine
50	Fifty
35	Thirty five
26	Twenty six
22	Twenty two
4	Four

NUMBERS 1-50 EXERCISE 40

9	Nine
47	Forty seven
41	Forty one
3	Three
18	Eighteen
21	Twenty one
26	Twenty six
43	Forty three
28	Twenty eight
14	Fourteen

NUMBERS 1-50 EXERCISE 41

47 Forty seven
48 Forty eight
39 Thirty nine
29 Twenty nine
28 Twenty eight
45 Forty five
38 Thirty eight
23 Twenty three
35 Thirty five
18 Eighteen

NUMBERS 1-50 EXERCISE 42

18   Eighteen
44   Forty four
9   Nine
27   Twenty seven
48   Forty eight
32   Thirty two
24   Twenty four
10   Ten
38   Thirty eight
45   Forty five

NUMBERS 1-50 EXERCISE 43

43	Forty three
24	Twenty four
35	Thirty five
46	Forty six
9	Nine
22	Twenty two
20	Twenty
18	Eighteen
21	Twenty one
8	Eight

NUMBERS 1-50 EXERCISE 44

44	Forty four
16	Sixteen
14	Fourteen
27	Twenty seven
25	Twenty five
50	Fifty
1	One
11	Eleven
48	Forty eight
33	Thirty three

NUMBERS 1-50 EXERCISE 45

24	Twenty four
11	Eleven
26	Twenty six
36	Thirty six
39	Thirty nine
44	Forty four
43	Forty three
46	Forty six
50	Fifty
20	Twenty

NUMBERS 1-50 EXERCISE 46

27 Twenty seven

12 Twelve

24 Twenty four

17 Seventeen

49 Forty nine

33 Thirty three

9 Nine

34 Thirty four

47 Forty seven

45 Forty five

NUMBERS 1-50 EXERCISE 47

23	Twenty three
43	Forty three
34	Thirty four
50	Fifty
45	Forty five
4	Four
13	Thirteen
24	Twenty four
21	Twenty one
25	Twenty five
1	One
19	Nineteen

NUMBERS 1-50 EXERCISE 48

16 Sixteen
47 Forty seven
25 Twenty five
46 Forty six
3 Three
22 Twenty two
5 Five
11 Eleven
7 Seven
29 Twenty nine

NUMBERS 1-50 EXERCISE 49

35 Thirty five

50 Fifty

49 Forty nine

4 Four

31 Thirty one

33 Thirty three

12 Twelve

13 Thirteen

14 Fourteen

17 Seventeen

NUMBERS 1-50 EXERCISE 50

13	Thirteen
35	Thirty Five
7	Seven
19	Nineteen
10	Ten
32	Thirty two
18	Eighteen
47	Forty seven
4	Four
24	Twenty four

NUMBERS 1-50 EXERCISE 51

28	Twenty eight
37	Thirty seven
39	Thirty nine
13	Thirteen
20	Twenty
6	Six
10	Ten
19	Nineteen
36	Thirty six
5	Five

NUMBERS 1-50 EXERCISE 52

49	Forty nine
43	Forty three
9	Nine
23	Twenty three
22	Twenty two
34	Thirty four
17	Seventeen
21	Twenty one
18	Eighteen
32	Thirty two

NUMBERS 1-50 EXERCISE 53

14 Fourteen
37 Thirty seven
7   Seven
36 Thirty six
4   Four
24 Twenty four
41 Forty one
16 Sixteen
33 Thirty three
44 Forty four

NUMBERS 1-50 EXERCISE 54

37 Thirty seven

15 Fifteen

14 Fourteen

31 Thirty one

17 Seventeen

30 Thirty

23 Twenty three

12 Twelve

10 Ten

4 Four

NUMBERS 1-50 EXERCISE 55

18 Eighteen
41 Forty-one
20 Twenty
42 Forty-two
9 Nine
31 Thirty-one
29 Twenty-nine
39 Thirty-nine
6 Six
35 Thirty-five

NUMBERS 1-50 EXERCISE 56

29 Twenty nine
17 Seventeen
18 Eighteen
39 Thirty nine
50 Fifty
40 Forty
20 Twenty
49 Forty nine
44 Forty four
42 Forty two

NUMBERS 1-50 EXERCISE 57

48	Forty eight
33	Thirty three
43	Forty three
27	Twenty seven
4	Four
18	Eighteen
39	Thirty nine
42	Forty two
7	Seven
29	Twenty nine

NUMBERS 1-50 EXERCISE 58

17 Seventeen
33 Thirty three
39 Thirty nine
7   Seven
18 Eighteen
37 Thirty seven
30 Thirty
8   Eight
1   One
48 Forty eight
10 Ten
2   Two

NUMBERS 1-50 EXERCISE 59

8   Eight

35   Thirty five

28   Twenty eight

25   Twenty five

11   Eleven

44   Forty four

9   Nine

26   Twenty six

12   Twelve

47   Forty seven

NUMBERS 1-50 EXERCISE 60

29 Twenty nine
31 Thirty one
14 Fourteen
45 Forty five
24 Twenty four
25 Twenty five
41 Forty one
36 Thirty six
32 Thirty two
21 Twenty one

NUMBERS 1-50 EXERCISE 61

8 — Eight
19 — Nineteen
25 — Twenty five
21 — Twenty one
39 — Thirty nine
24 — Twenty four
3 — Three
46 — Forty six
31 — Thirty one
22 — Twenty two

NUMBERS 1-50 EXERCISE 62

19	Nineteen
14	Fourteen
44	Forty four
6	Six
18	Eighteen
9	Nine
43	Forty three
16	Sixteen
36	Thirty six
25	Twenty five

NUMBERS 1-50 EXERCISE 63

9   Nine

6   Six

49   Forty nine

20   Twenty

18   Eighteen

11   Eleven

5   Five

13   Thirteen

31   Thirty one

41   Forty one

NUMBERS 1-50 EXERCISE 64

11 Eleven
32 Thirty two
26 Twenty six
29 Twenty nine
39 Thirty nine
38 Thirty eight
3 Three
34 Thirty four
30 Thirty
9 Nine

NUMBERS 1-50 EXERCISE 65

2 — Two
23 — Twenty three
37 — Thirty seven
25 — Twenty five
38 — Thirty eight
1 — One
4 — Four
14 — Fourteen
47 — Forty seven
7 — Seven

NUMBERS 1-50 EXERCISE 66

7 — Seven
35 — Thirty five
23 — Twenty three
31 — Thirty one
39 — Thirty nine
25 — Twenty five
12 — Twelve
50 — Fifty
47 — Forty seven
32 — Thirty two

NUMBERS 1-50 EXERCISE 67

15	Fifteen
26	Twenty six
30	Thirty
5	Five
29	Twenty nine
32	Thirty two
28	Twenty eight
33	Thirty three
10	Ten
44	Forty four

NUMBERS 1-50 EXERCISE 68

28	Twenty eight
15	Fifteen
30	Thirty
5	Five
26	Twenty six
47	Forty seven
18	Eighteen
45	Forty five
8	Eight
13	Thirteen

NUMBERS 1-50 EXERCISE 69

11	Eleven
25	Twenty five
49	Forty nine
6	Six
33	Thirty three
3	Three
15	Fifteen
39	Thirty nine
47	Forty seven
45	Forty five
38	Thirty eight
48	Forty eight

NUMBERS 1-50 EXERCISE 70

14 Fourteen
47 Forty seven
12 Twelve
24 Twenty four
32 Thirty two
46 Forty six
22 Twenty two
34 Thirty four
33 Thirty three
6 Six

NUMBERS 1-50 EXERCISE 71

6   Six

25   Twenty five

20   Twenty

23   Twenty three

4   Four

11   Eleven

34   Thirty four

40   Forty

14   Fourteen

47   Forty seven

NUMBERS 1-50 EXERCISE 72

21 Twenty one
8 Eight
15 Fifteen
41 Forty one
36 Thirty six
49 Forty nine
1 One
19 Nineteen
5 Five
47 Forty seven

NUMBERS 1-50 EXERCISE 73

43 Forty three

3 Three

26 Twenty six

2 Two

11 Eleven

9 Nine

37 Thirty seven

47 Forty seven

33 Thirty three

21 Twenty one

NUMBERS 1-50 EXERCISE 74

3	Three
36	Thirty six
18	Eighteen
42	Forty two
34	Thirty four
47	Forty seven
40	Forty
23	Twenty three
12	Twelve
21	Twenty one

NUMBERS 1-50 EXERCISE 75

8	Eight
34	Thirty four
45	Forty five
2	Two
28	Twenty eight
46	Forty six
1	One
16	Sixteen
7	Seven
11	Eleven

NUMBERS 1-50 EXERCISE 76

11	Eleven
48	Forty eight
18	Eighteen
12	Twelve
28	Twenty eight
14	Fourteen
37	Thirty seven
40	Forty
1	One
21	Twenty one

NUMBERS 1-50 EXERCISE 77

48	Forty eight
18	Eighteen
24	Twenty four
42	Forty two
27	Twenty seven
11	Eleven
19	Nineteen
45	Forty five
14	Fourteen
31	Thirty one

NUMBERS 1-50 EXERCISE 78

15 Fifteen
27 Twenty seven
25 Twenty five
24 Twenty four
8 Eight
36 Thirty six
46 Forty six
11 Eleven
50 Fifty
45 Forty five

NUMBERS 1-50 EXERCISE 79

35	Thirty five
13	Thirteen
16	Sixteen
42	Forty two
15	Fifteen
41	Forty one
39	Thirty nine
24	Twenty four
4	Four
43	Forty three

NUMBERS 1-50 EXERCISE 80

45	Forty five
7	Seven
31	Thirty one
37	Thirty seven
11	Eleven
46	Forty six
18	Eighteen
28	Twenty eight
43	Forty three
42	Forty two
15	Fifteen
12	Twelve

NUMBERS 1-50 EXERCISE 81

48	Forty eight
8	Eight
2	Two
41	Forty one
20	Twenty
18	Eighteen
7	Seven
39	Thirty nine
50	Fifty
19	Nineteen

NUMBERS 1-50 EXERCISE 82

5   Five

33   Thirty three

13   Thirteen

38   Thirty eight

44   Forty four

14   Fourteen

37   Thirty seven

25   Twenty five

21   Twenty one

18   Eighteen

NUMBERS 1-50 EXERCISE 83

1   One
40   Forty
15   Fifteen
35   Thirty five
26   Twenty six
45   Forty five
19   Nineteen
29   Twenty nine
32   Thirty two
36   Thirty six

NUMBERS 1-50 EXERCISE 84

23	Twenty three
27	Twenty seven
31	Thirty one
45	Forty five
5	Five
22	Twenty two
42	Forty two
38	Thirty eight
16	Sixteen
32	Thirty two

NUMBERS 1-50 EXERCISE 85

9 — Nine
48 — Forty eight
6 — Six
19 — Nineteen
44 — Forty four
12 — Twelve
26 — Twenty six
35 — Thirty five
30 — Thirty
15 — Fifteen

NUMBERS 1-50 EXERCISE 86

49 Forty nine
19 Nineteen
6 Six
15 Fifteen
2 Two
31 Thirty one
21 Twenty one
29 Twenty nine
45 Forty five
14 Fourteen

NUMBERS 1-50 EXERCISE 87

43	Forty three
41	Forty one
23	Twenty three
3	Three
7	Seven
2	Two
46	Forty six
30	Thirty
50	Fifty
13	Thirteen

NUMBERS 1-50 EXERCISE 88

48 Forty eight
17 Seventeen
7 Seven
18 Eighteen
25 Twenty five
34 Thirty four
26 Twenty six
11 Eleven
6 Six
14 Fourteen

NUMBERS 1-50 EXERCISE 89

12	Twelve
37	Thirty seven
20	Twenty
10	Ten
45	Forty five
4	Four
16	Sixteen
15	Fifteen
46	Forty six
19	Nineteen

NUMBERS 1-50 EXERCISE 90

8	Eight
31	Thirty one
29	Twenty nine
18	Eighteen
48	Forty eight
26	Twenty six
10	Ten
43	Forty three
36	Thirty six
37	Thirty seven

TO SAY THANK YOU FOR PURCHASING THIS BOOK I AM GIVING AWAY A FREE COPY OF MY LATEST CHILDREN'S BOOK OR PUZZLE BOOK!

JUST GO TO THE LINK BELOW AND COMMENT WITH SITE LINK FOR BOOK REVIEW TO RECEIVE YOUR COPY

http://sudokuprintable.blogspot.com

CHECK OUT MORE BOOKS BELOW

All book collections can be found on

https://www.createspace.com/7511563
https://www.createspace.com/7512265
https://www.createspace.com/7512005

YOU MIGHT ALSO BE INTERESTED IN MY NEWEST COLLECTION OF BOOKS

CHECK OUT THE BOOKS ON THE NEXT PAGE

500 SUDOKU PUZZLES WITH ANSWERS

FIND OUT MORE HERE:

https://tinyurl.com/sudoku501

www.ingramcontent.com/pod-product-compliance
Lightning Source LLC
Chambersburg PA
CBHW082211220526
45470CB00010B/3123